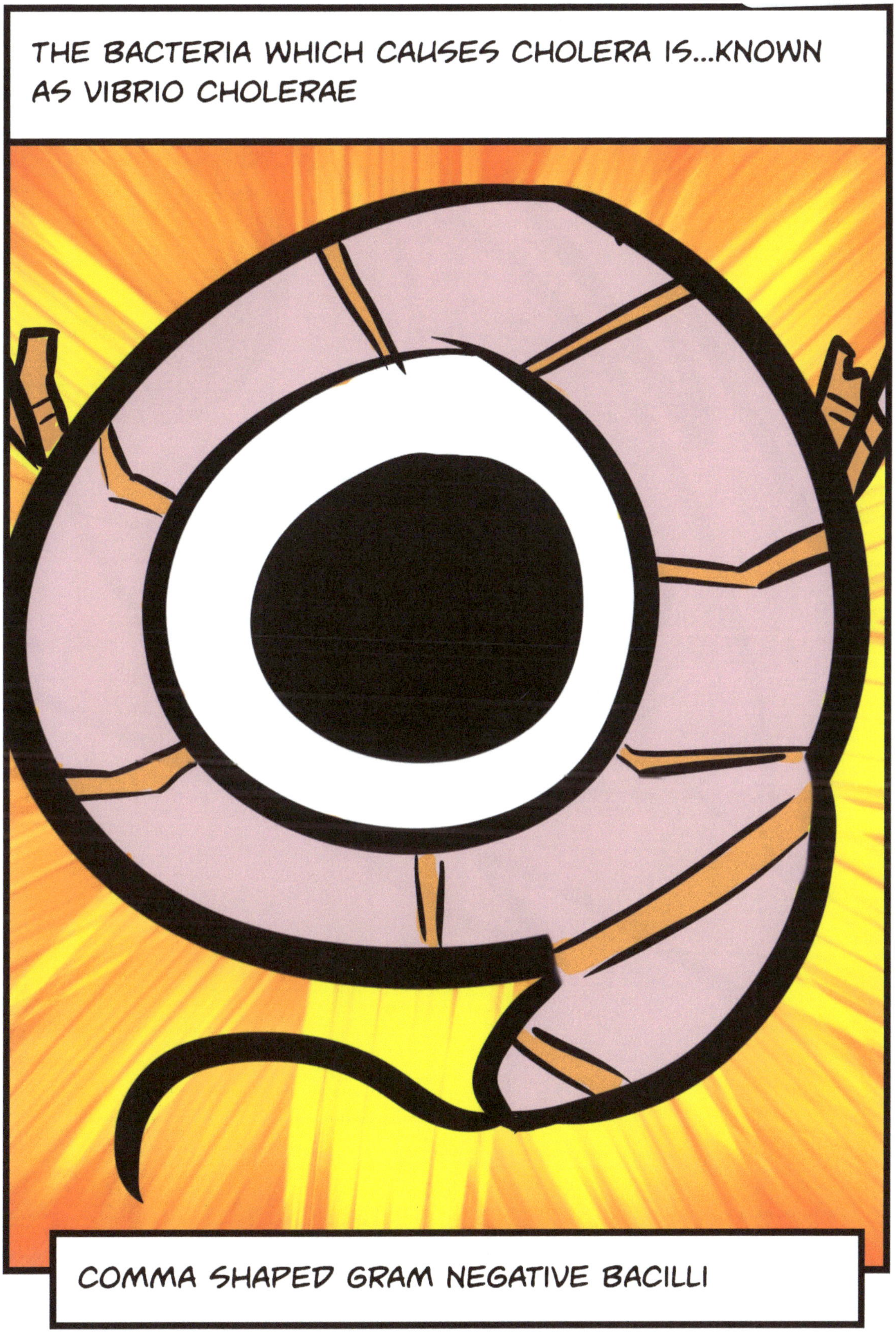
THE BACTERIA WHICH CAUSES CHOLERA IS...KNOWN AS VIBRIO CHOLERAE
COMMA SHAPED GRAM NEGATIVE BACILLI

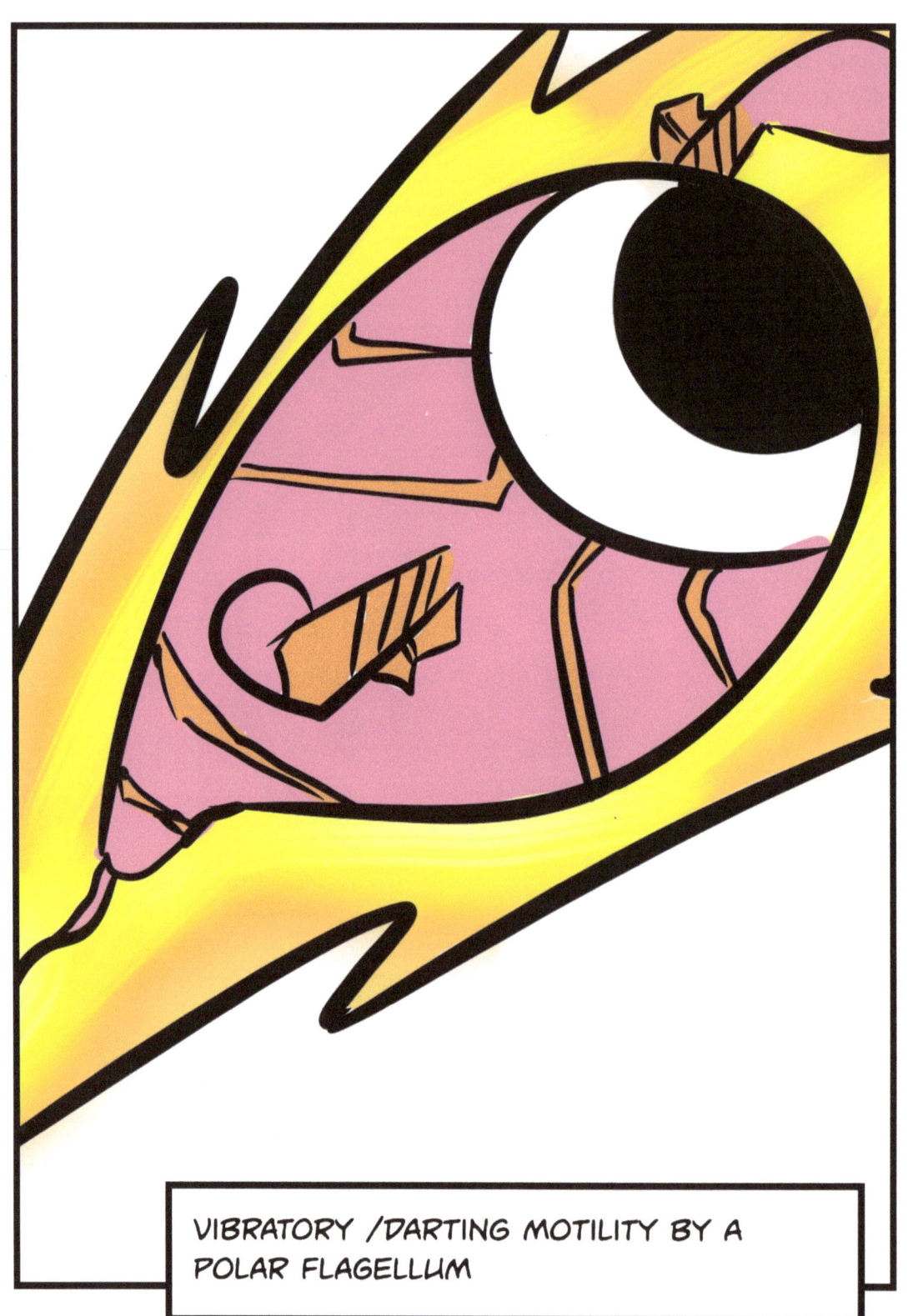

VIBRATORY /DARTING MOTILITY BY A POLAR FLAGELLUM

CHOLERA

> SEVENTH PANDEMIC WAS CAUSED BY ELTOR BIOTYPE AND ORIGINATED IN INDONESIA. IT WAS MILDER BUT HAD HIGH CARRIER RATE.

CHOLERA IS EXCLUSIVELY HUMAN DISEASE.

SOURCE OF INFECTION IS PATIENT OR CARRIER.

"O" BLOOD GROUP INDIVIDUALS ARE MORE SUSCEPTIBLE THAN "AB" BLOOD GROUP

INFECTION IS ACQUIRED BY INGESTION OF FECALLY CONTAMINATED FOOD OR WATER.

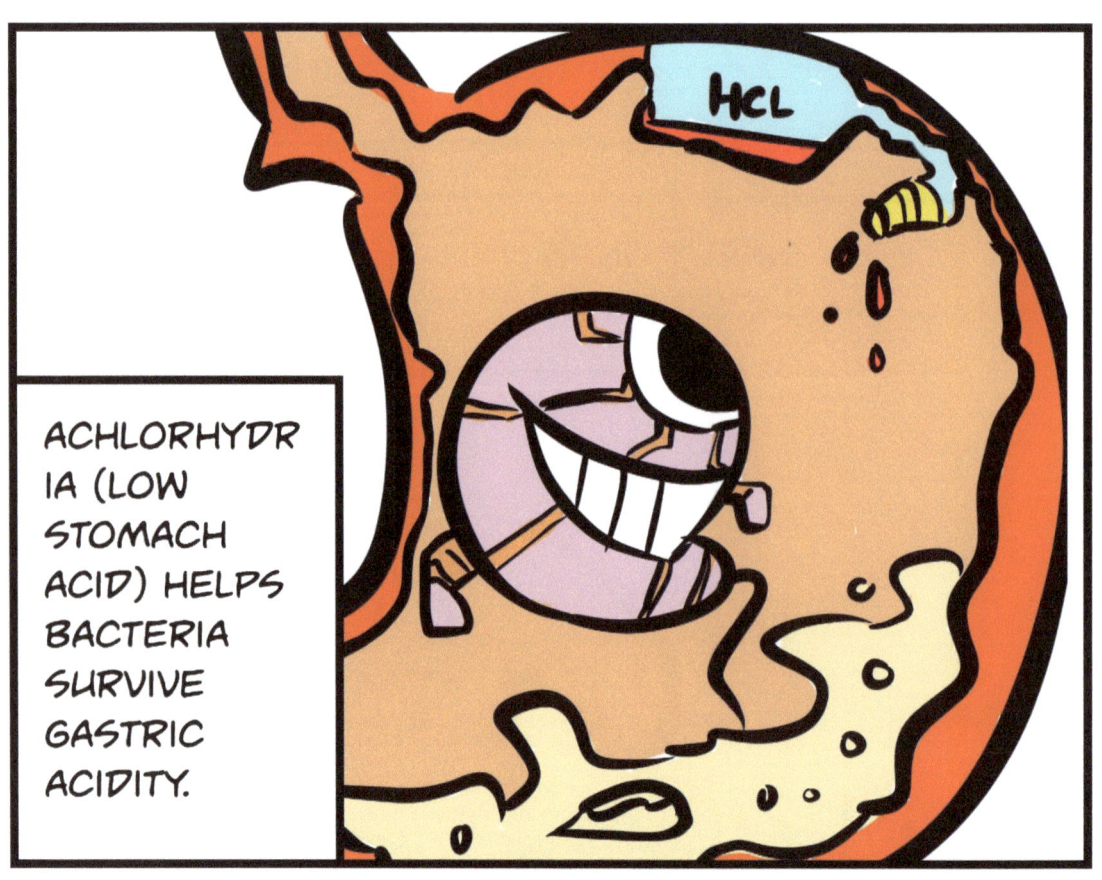

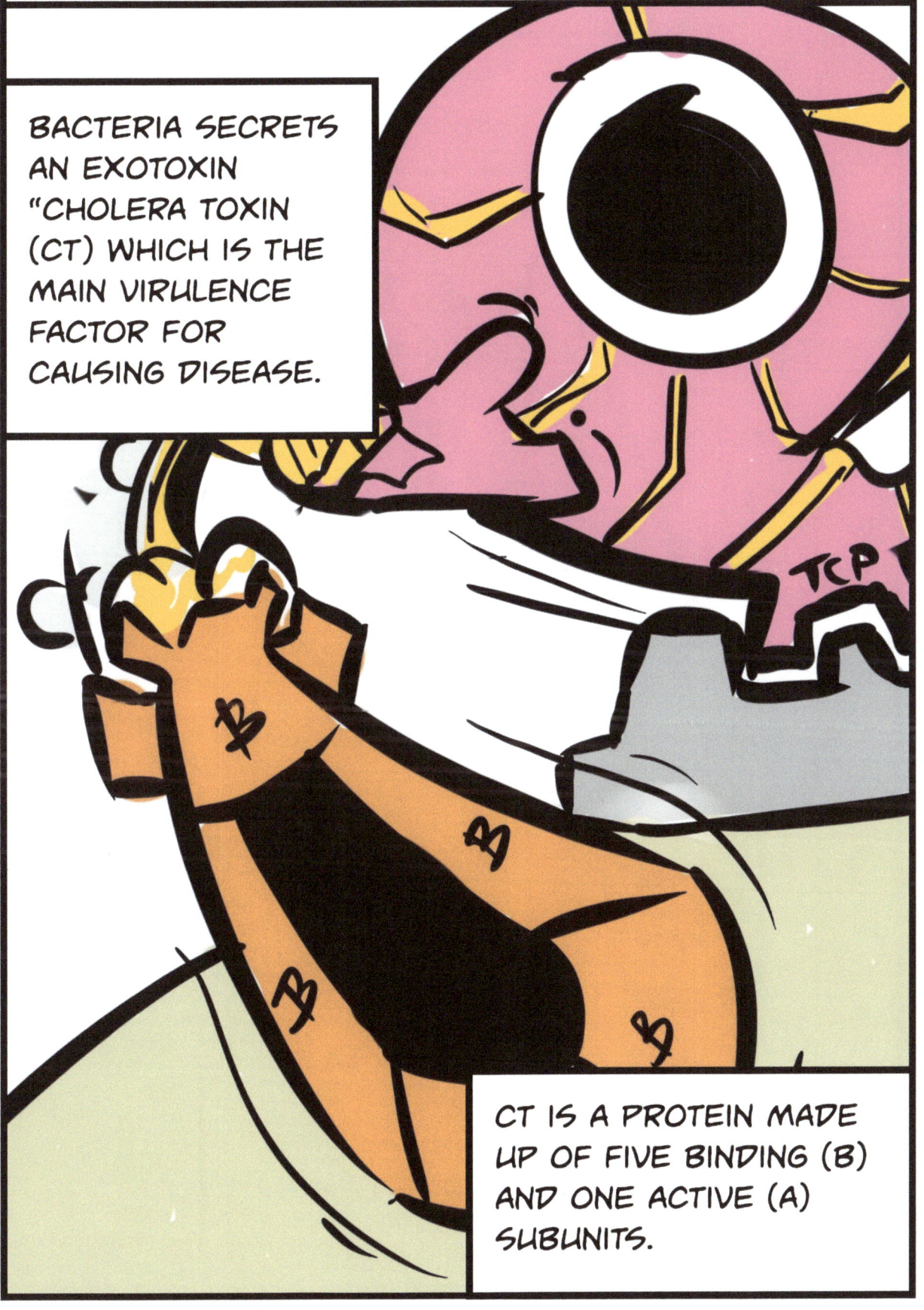

CT BINDING SUBUNITS BIND TO GM1 GANGLIOSIDE RECEPTORS ON SMALL INTESTINAL EPITHELIAL CELLS.

A SUBUNIT IS INTERNALISED AND SPLIT INTO TWO FRAGMENTS A1 AND A2.

A1 SUBUNIT CAUSES ADP RIBOSYLATION OF ADENYL CYCLASE ENZYME AND CAUSE CONTINUOUS STIMULATION OF CAMP PRODUCTION.

CAMP CAUSES WATER AND ELECTROLYTES TO MOVE OUT OF CELLS INTO INTESTINAL LUMEN WHICH LEADS TO WATERY DIARRHOEA.

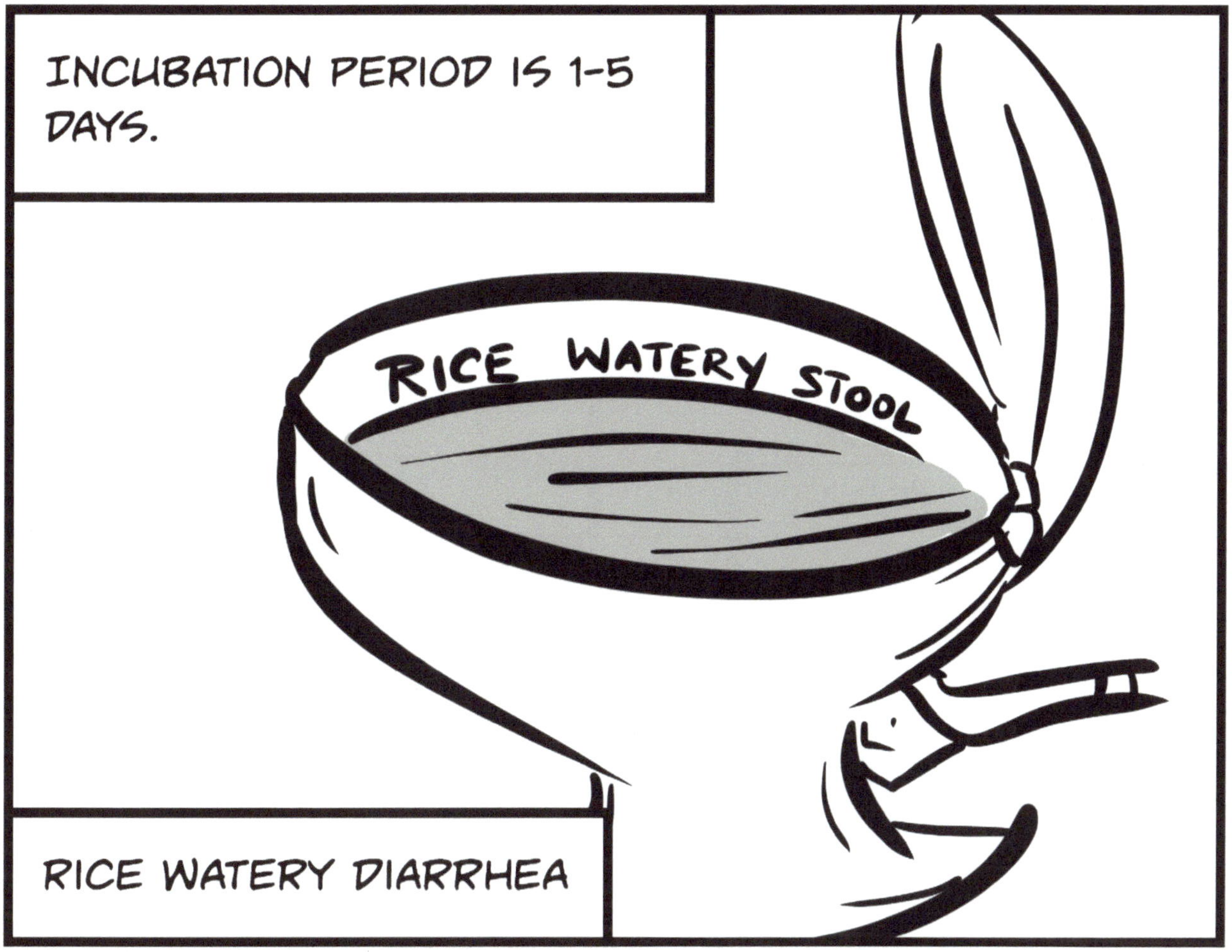
INCUBATION PERIOD IS 1-5 DAYS.
RICE WATERY STOOL
RICE WATERY DIARRHEA

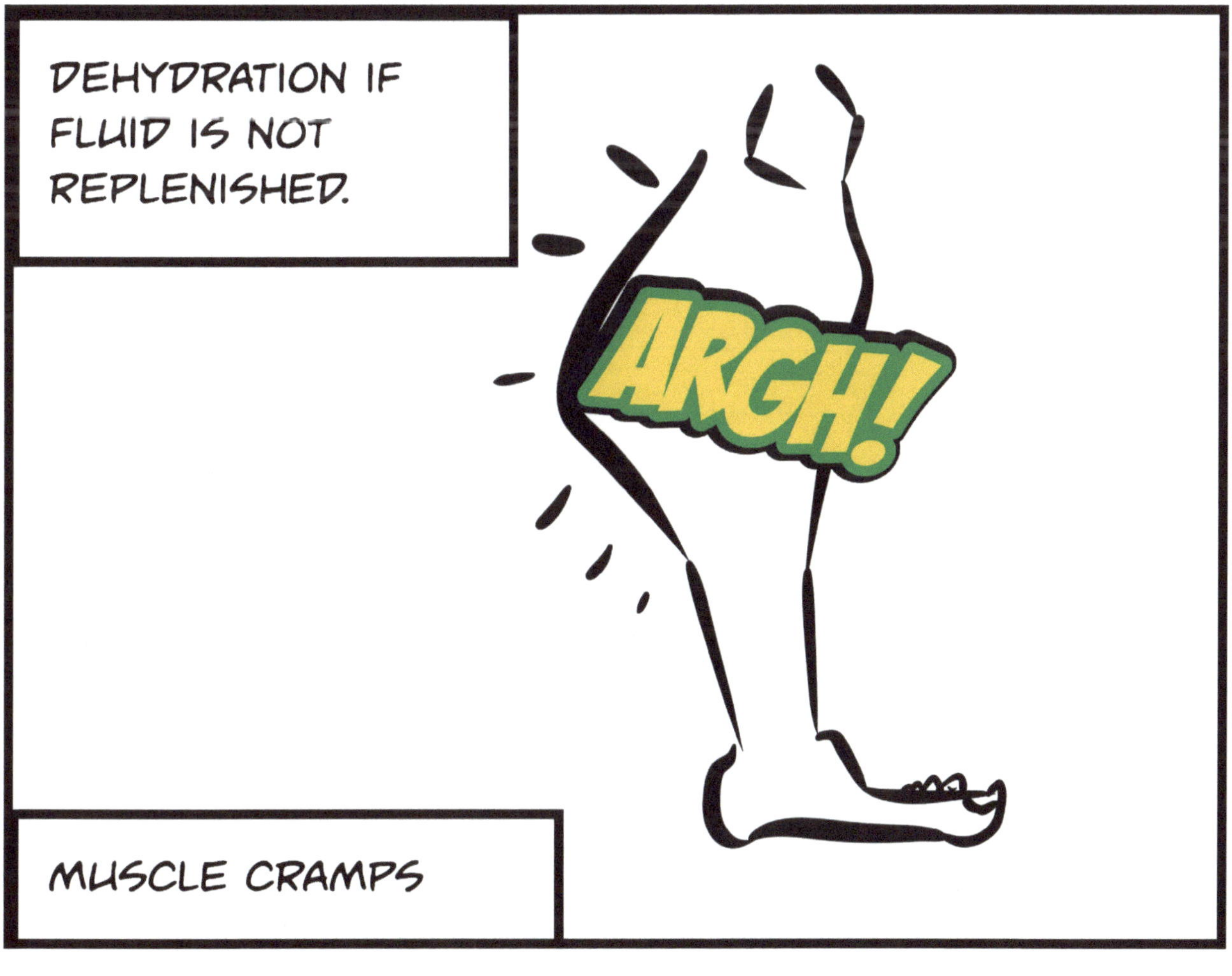
DEHYDRATION IF FLUID IS NOT REPLENISHED.
ARGH!
MUSCLE CRAMPS

LAB DIAGNOSIS

STOOL AND RECTAL SWABS ARE COLLECTED FOR LAB DIAGNOSIS OF CHOLERA.

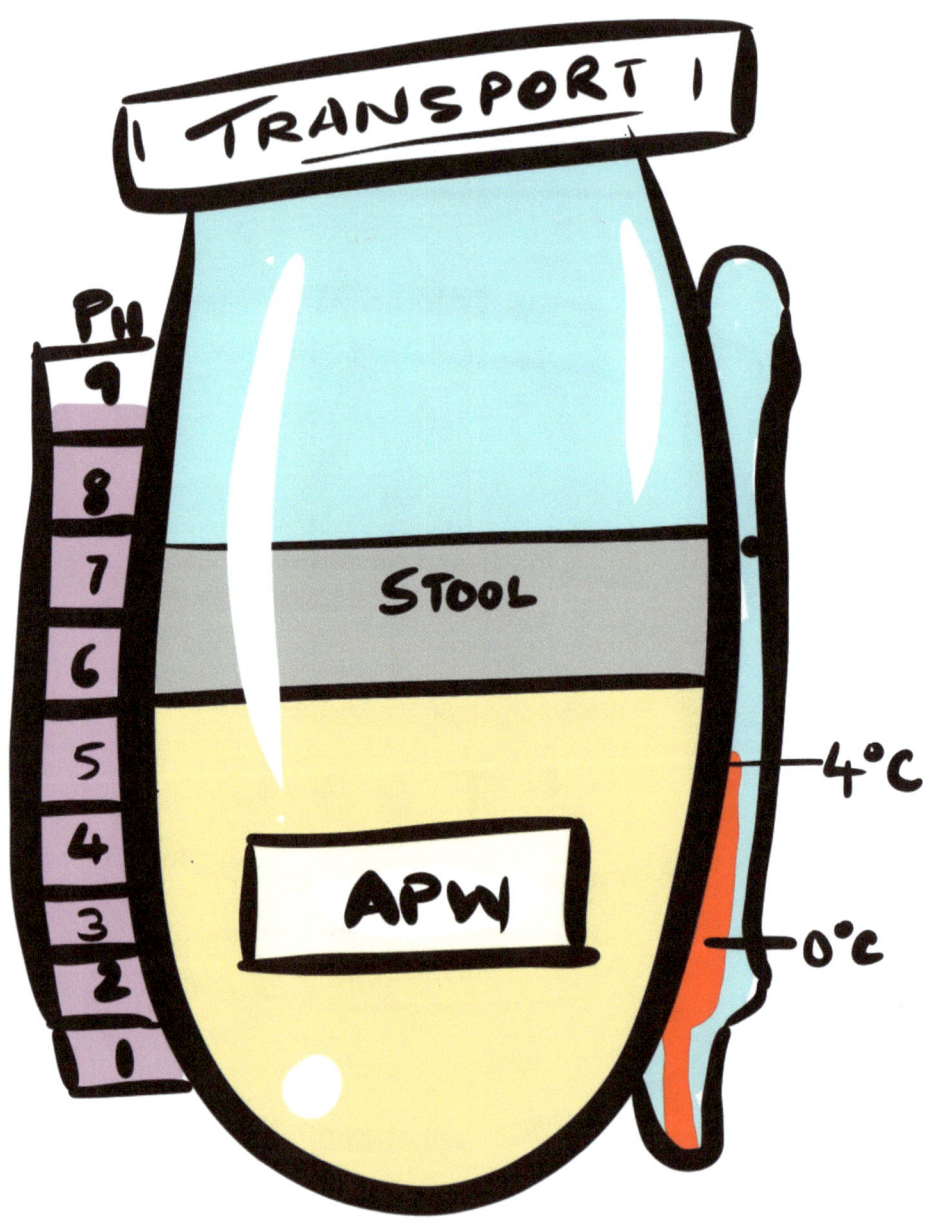

IF LONGER DELAY IS EXPECTED IN TRANSPORTATION....

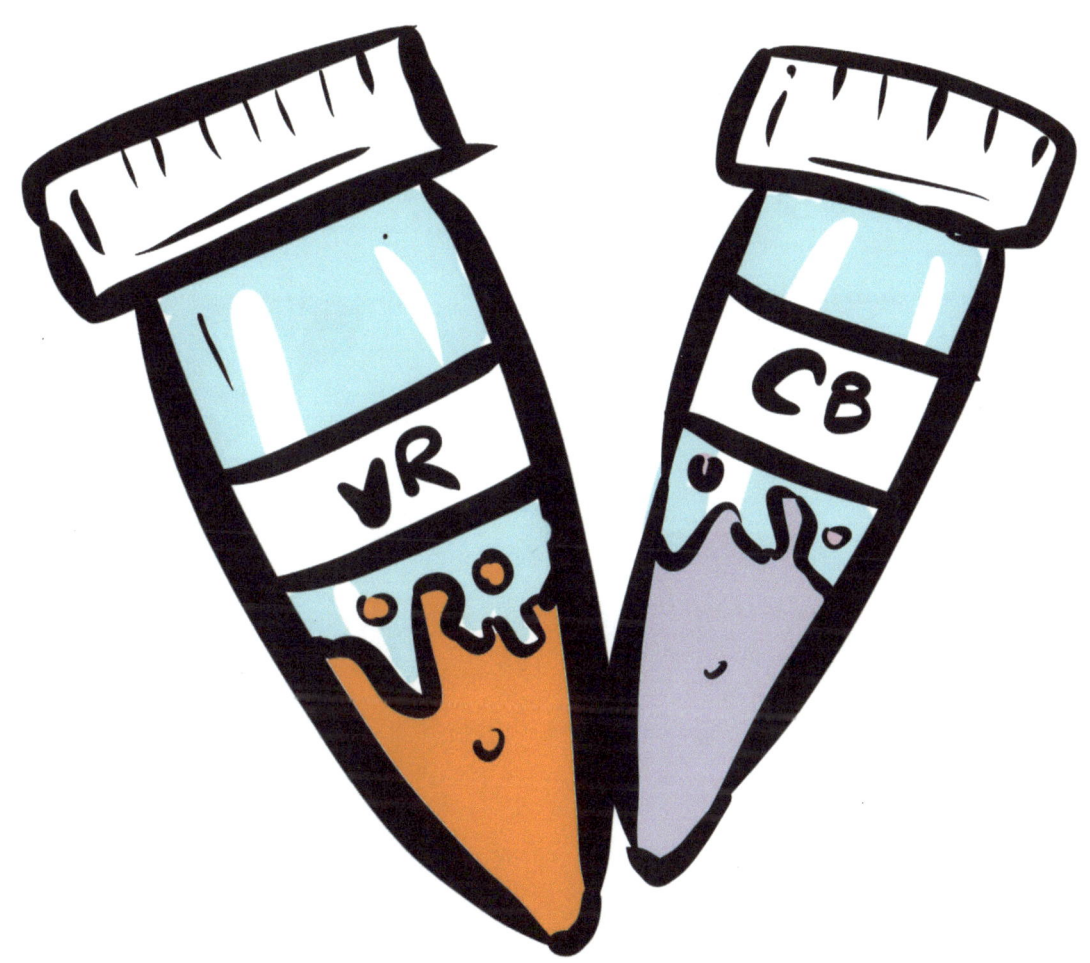

THEN VR (VENKATRAMAN RAMKRISHNAN) OR CB (CARRY BLAIR) MEDIUM IS TO BE USED.

OTHER OPTIONS ARE ENRICHMENT AND TRANSPORT MEDIA LIKE...

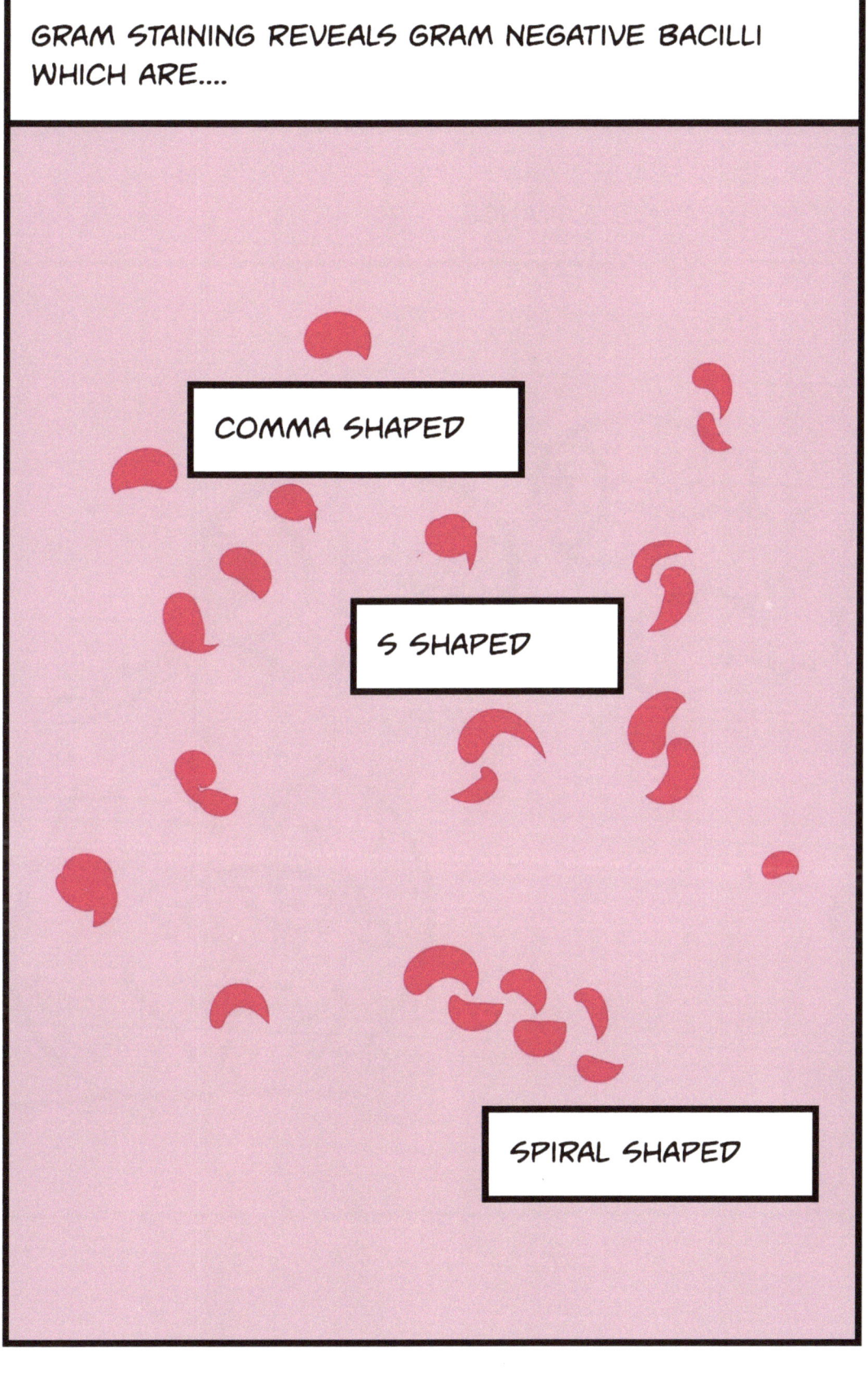

6-8 HOURS ENRICHMENT IS RECOMMENDED BEFORE PLATING

```
┌─────────────────────────────┐
│      NUTRIENT AGAR          │
├─────────────────────────────┤
│      MACCONKEY AGAR         │
├─────────────────────────────┤
│      BLOOD AGAR             │
└─────────────────────────────┘
┌─────────────────────────────────┐
│   SELECTIVE MEDIUM: TCBS        │
└─────────────────────────────────┘
┌─────────────────────────────────────────┐
│ OTHER SELECTIVE MEDIA ARE BILE          │
│ SALT AGAR (BSA) AND GTTA.               │
└─────────────────────────────────────────┘
                 ↓
┌─────────────────────────────────────────────────┐
│ 37 C - AEROBE - PH 8.2 OVERNIGHT INCUBATION     │
└─────────────────────────────────────────────────┘
```

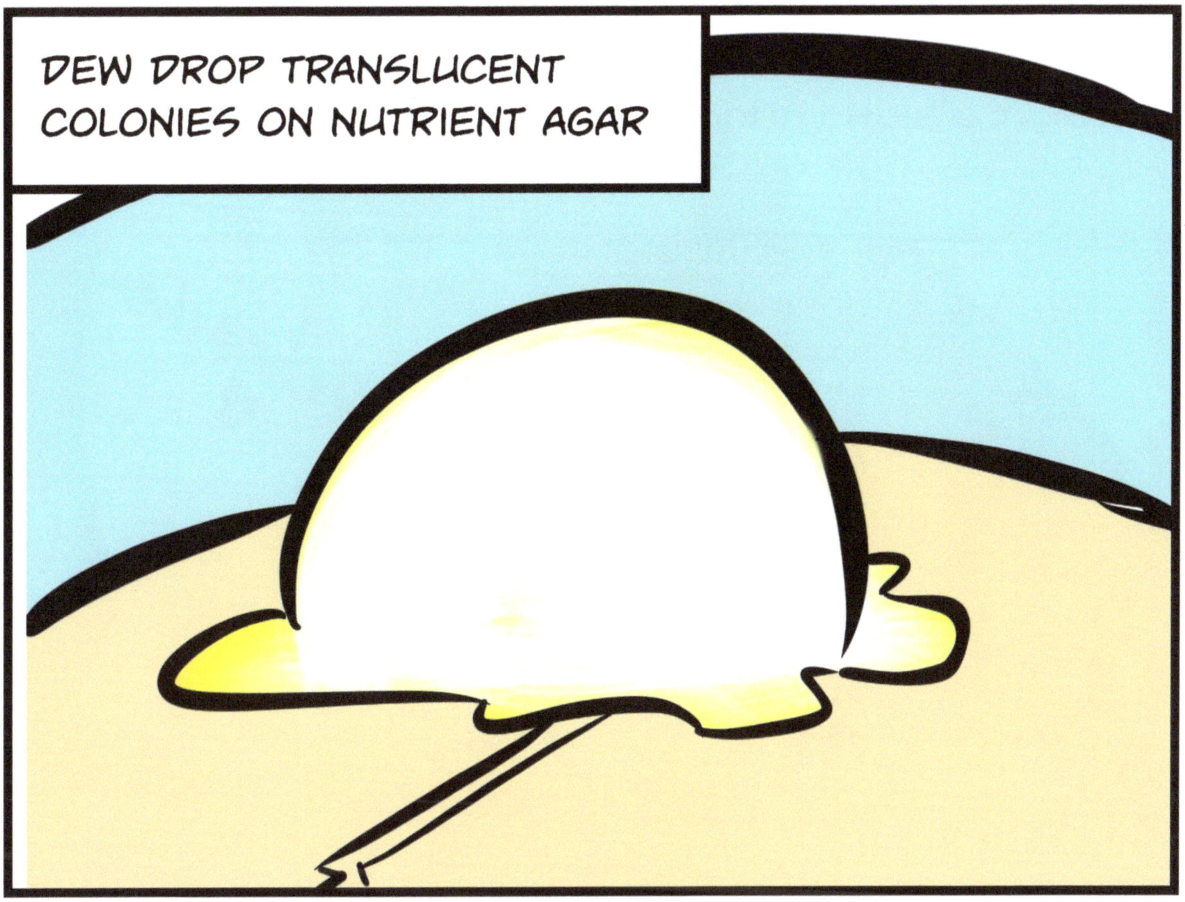
DEW DROP TRANSLUCENT COLONIES ON NUTRIENT AGAR

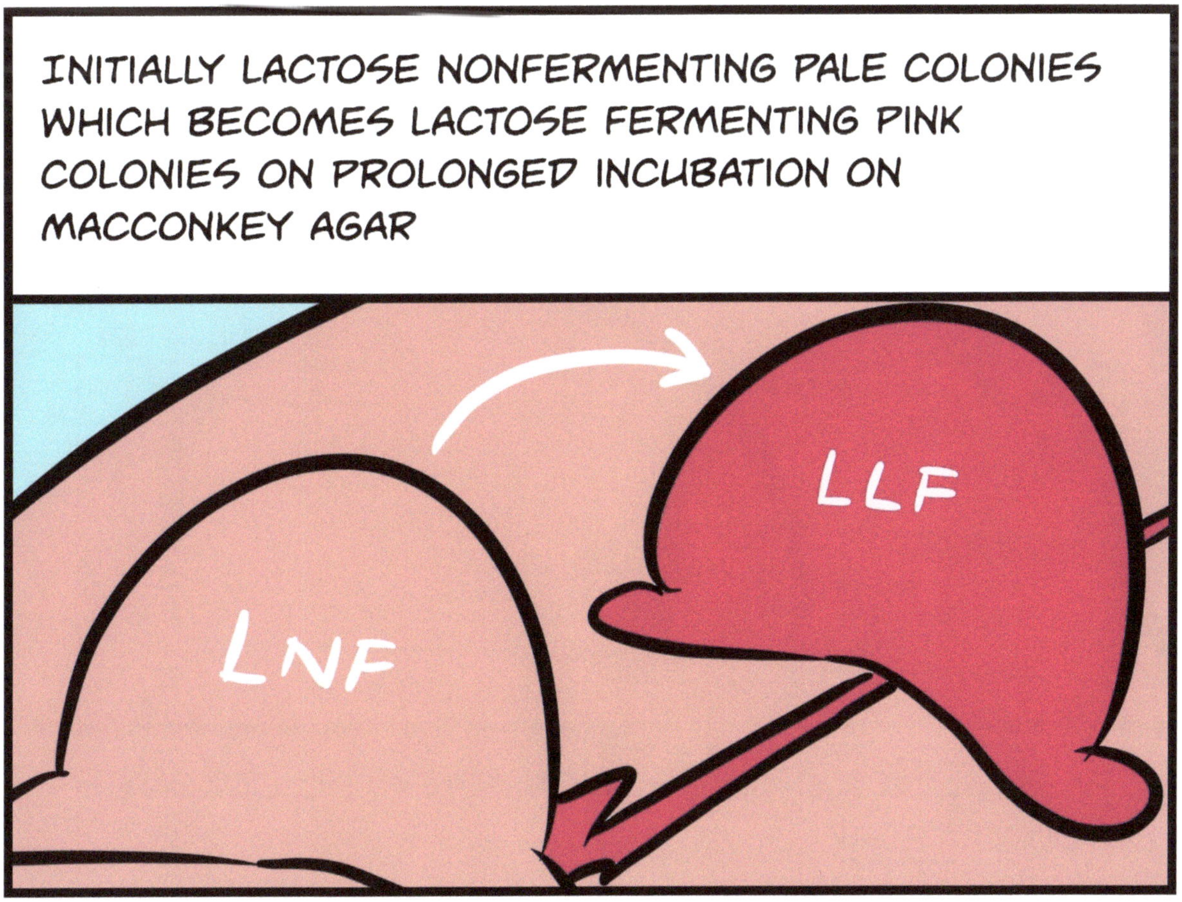
INITIALLY LACTOSE NONFERMENTING PALE COLONIES WHICH BECOMES LACTOSE FERMENTING PINK COLONIES ON PROLONGED INCUBATION ON MACCONKEY AGAR
LNF
LLF

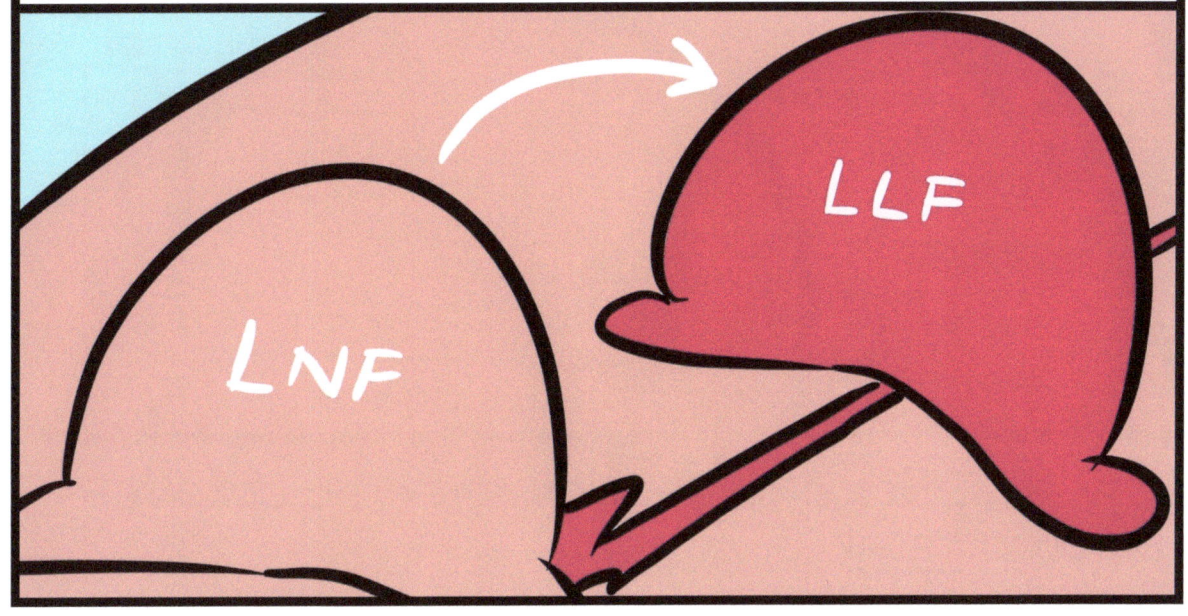

ELTOR VIBRIO GIVES ALPHA HEMOLYTIC COLONIES INITIALLY WHICH LATER ON SHOW CLEARING DUE TO HEMODIGESTION ON BLOOD AGAR

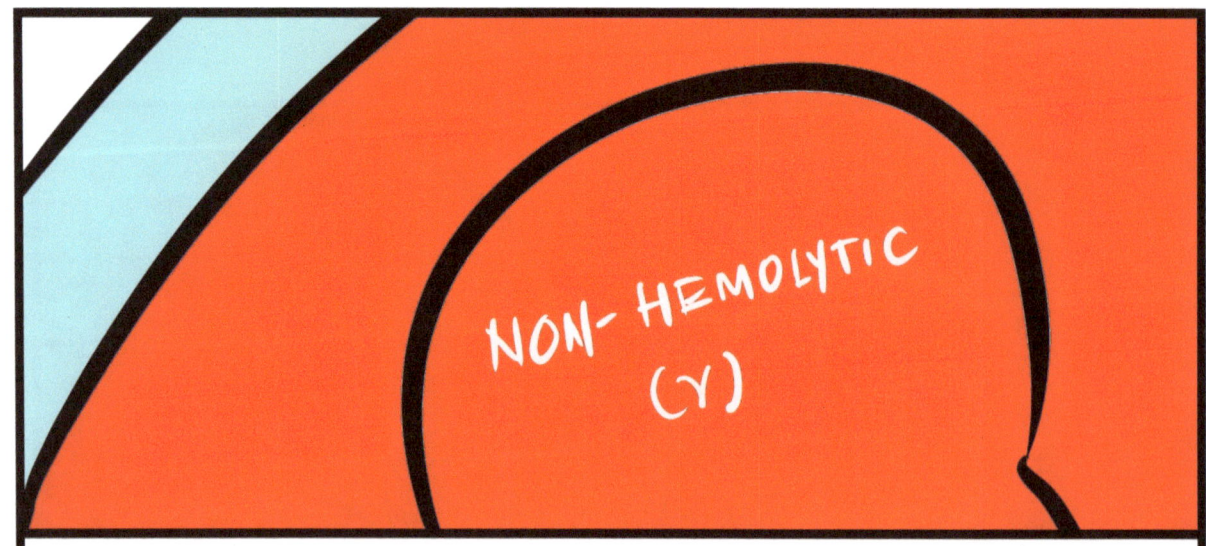

CLASSICAL VIBRIO SHOWS NONHEMOLYTIC COLONIES ON BLOOD AGAR.

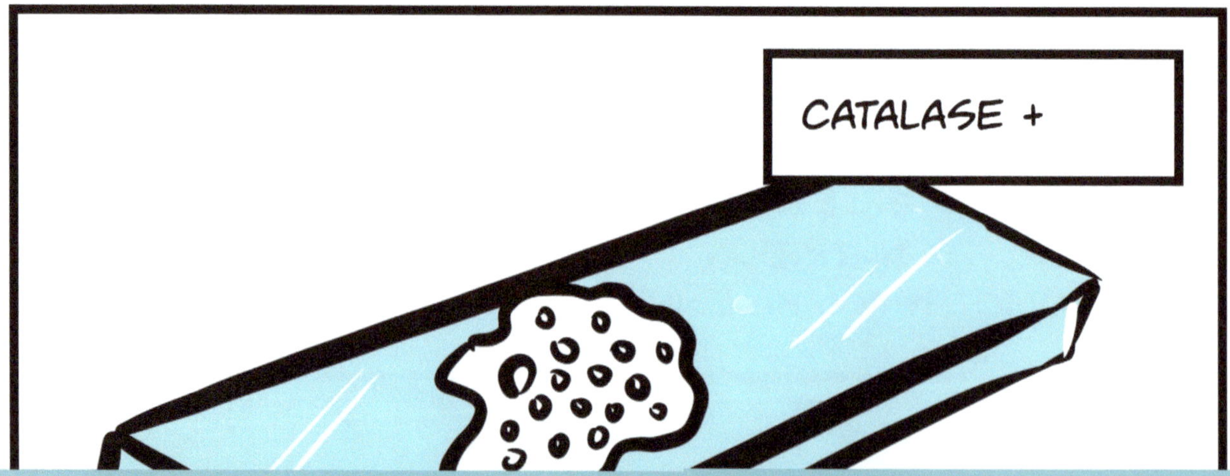
CATALASE +

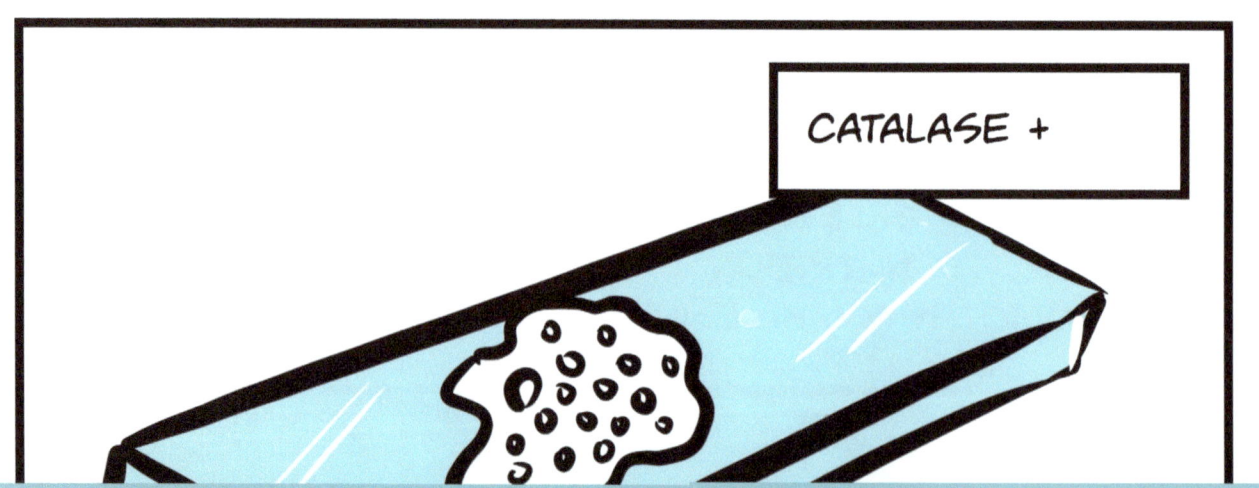

- FERMENT GLUCOSE, MALTOSE, MANNITOL AND SUCROSE WITH ACID ONLY. NO GAS PRODUCTION

- LATE LACTOSE FERMENTETER

- TSI : ACID/ ACID/ NO GAS / NO H2S

- GROW ONLY AT 0 % AND 1% NACL BROTH

- ORNITHINE, LYSINE, ARGININE : +/+/−

- INDOLE +

- NITRATE REDUCTION +

METHYL RED −

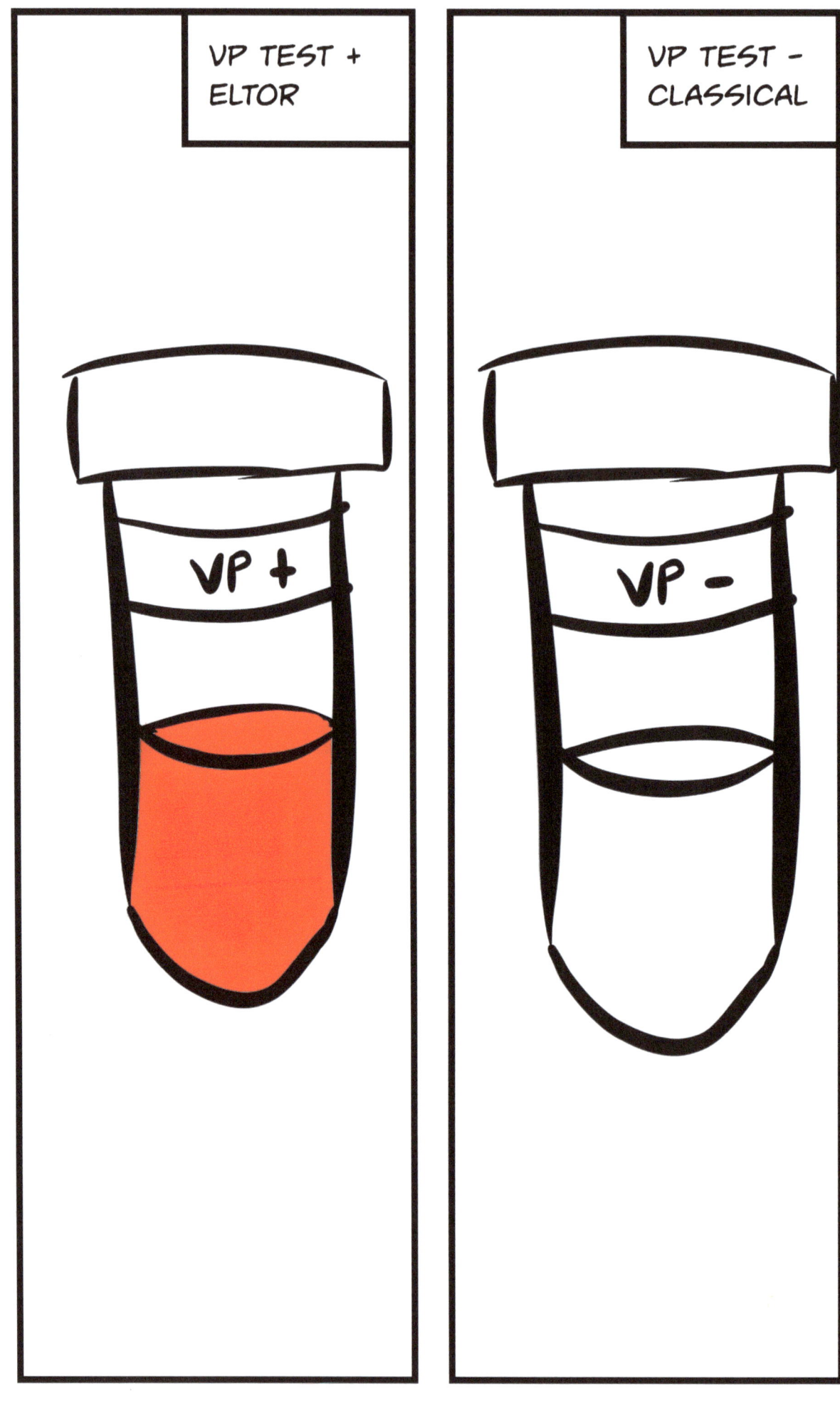

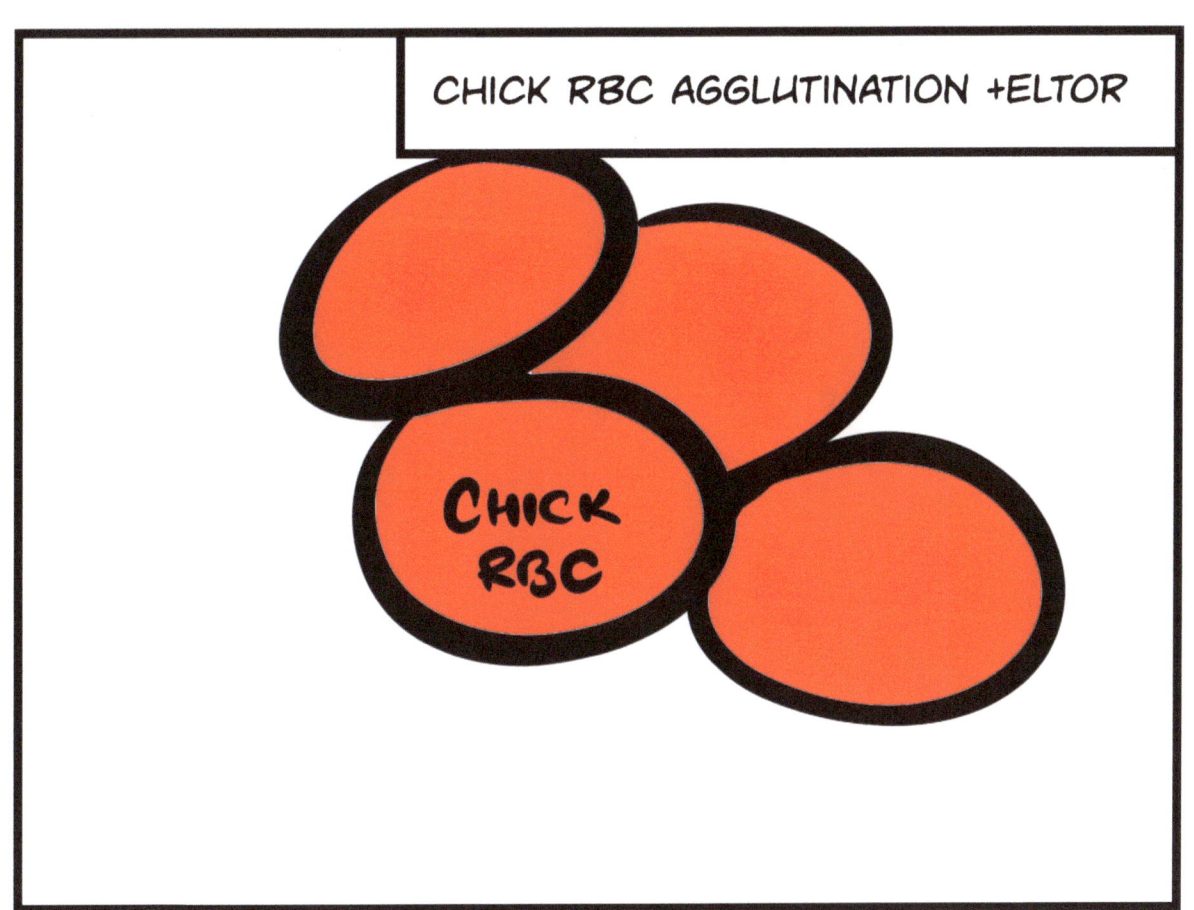
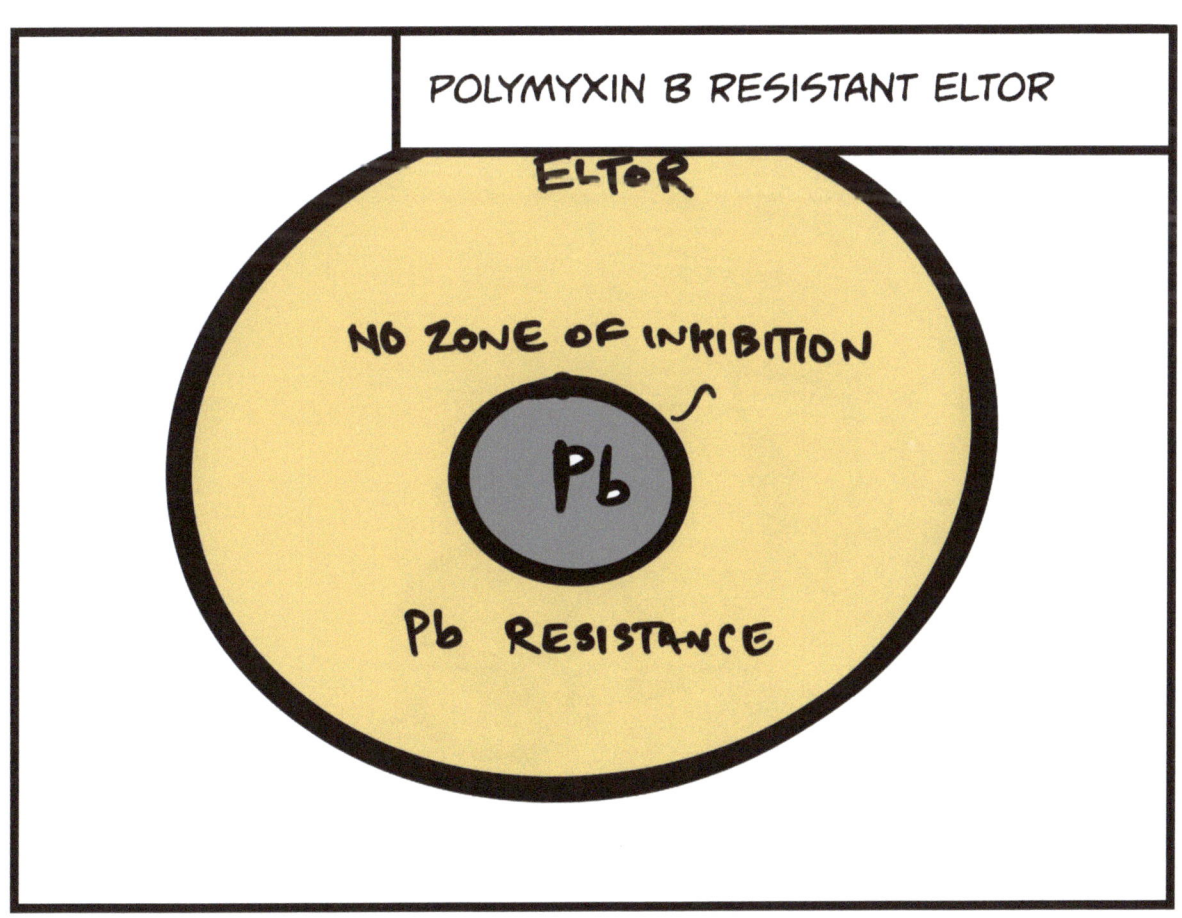

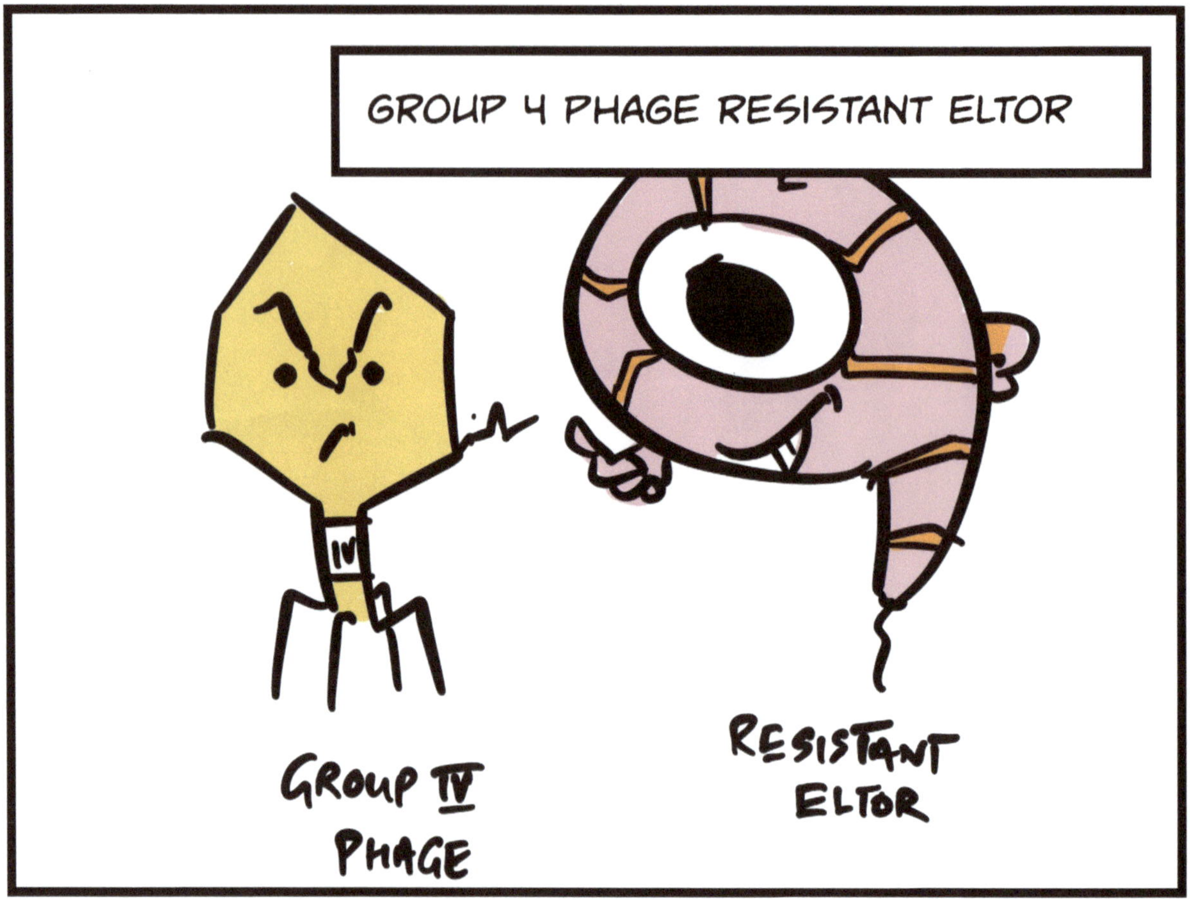
GROUP 4 PHAGE RESISTANT ELTOR
GROUP IV PHAGE
RESISTANT ELTOR

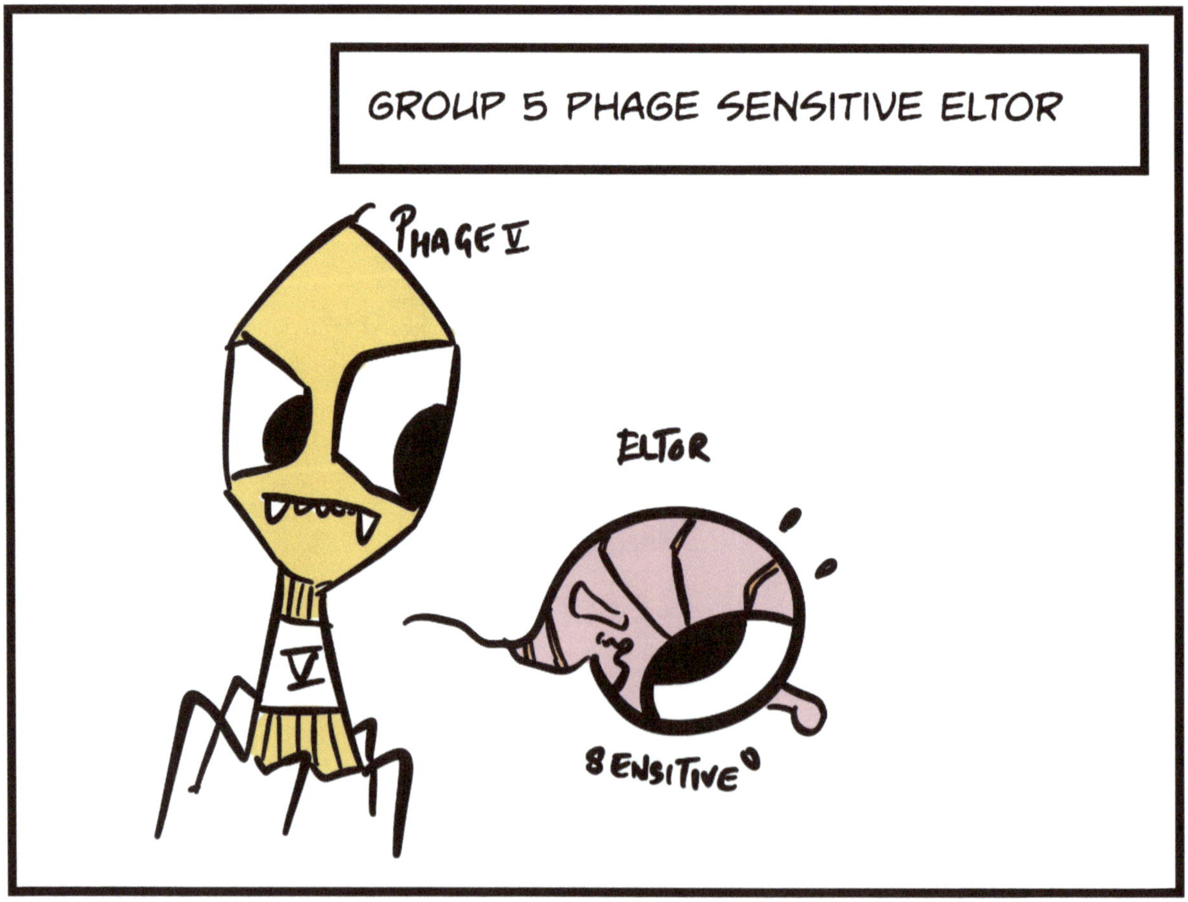
GROUP 5 PHAGE SENSITIVE ELTOR
PHAGE V
ELTOR
SENSITIVE

TREATMENT IS FLUID REPLACEMENT AND ANTIBIOTIC

KILLED CHOLERA VACCINE ARE AVAILABLE BUT ONLY OFFERS 50% PROTECTION RATE FOR 3-6 MONTHS. LIVE VACCINES ARE NOT SAFE. TOXOID VACCINE HAS BEEN NOT EFFECTIVE. RESEARCH FOR ORAL VACCINES ARE UNDER PROCESS.

www.ingramcontent.com/pod-product-compliance
Lightning Source LLC
Chambersburg PA
CBHW041307180526
45172CB00003B/998